Ah, the Beach!

Published by Willow Creek Press, Inc.
P.O. Box 147, Minocqua, Wisconsin 54548

Design: Donnie Rubo
Printed in China

Ah, the Beach!

SURF, SAND, SIMPLICITY

WILLOW CREEK PRESS®

The Sea, once it casts its spell,
holds one in its net of wonder forever.

—Jacques Cousteau

Why do we love the sea?
It is because it has some potent power
to make us think things
we like to think.

—Robert Henri

ONE LEARNS FIRST OF ALL
IN BEACH LIVING
THE ART OF SHEDDING;
HOW LITTLE ONE CAN
GET ALONG WITH,
NOT HOW
MUCH.

—Anne Morrow Lindbergh

The voice of the sea speaks to the soul.
The touch of the sea is sensuous,
enfolding the body in its soft,
close embrace.

—Kate Chopin

*If you're lucky enough
to be at the beach,
then you're lucky enough!*

—Unknown

One cannot collect all the
beautiful shells on the beach;
one can collect only a few,
and they are more beautiful if they are few.

—Anne Morrow Lindbergh

I could never stay long enough on the shore.
The tang of the untainted, fresh and free sea air
was like a cool, quieting thought.

—Helen Keller

IT'S THESE CHANGES IN LATITUDES

CHANGES IN ATTITUDES

NOTHING REMAINS QUITE THE SAME

WITH ALL OF OUR RUNNING

AND ALL OF OUR CUNNING

IF WE COULDN'T LAUGH

WE WOULD ALL

GO INSANE

—Jimmy Buffett, Changes in Latitudes, Changes in Attitude

I am you; you are ME.
You are the waves;
I am the ocean.
Know this and be free,

be divine.

—Sri Sathya Sai Baba

The seagull sees farthest who flies highest.

—*proverb*

*Eternity begins and ends
with the ocean's tides.*

—*Unknown*

LIVE IN THE SUNSHINE, SWIM THE SEA, DRINK THE WILD AIR.

—Ralph Waldo Emerson

From the east to the west
sped the angels of the Dawn,
from sea to sea, scattering light
with both their hands.

—H. Rider Haggard

The three great elemental sounds in nature
are the sound of rain,
the sound of wind in a primeval wood,
and the sound of outer ocean
on a beach.

—Henry Beston

Our memories of the ocean will linger on long after our footprints in the sand are gone.

—Unknown

I honestly think the beach is the only place children actually entertain themselves.

—Donna McLavy

NIBBLIN' ON SPONGE CAKES WATCHIN' THE SUN BAKE ALL OF THOSE TOURISTS COVERED IN OIL STRUMMIN' MY SIX ON MY FRONT PORCH SWING STRING SMELL THOSE SHRIMP THEY'RE BEGINNING TO BOIL

—Jimmy Buffett, *Margaritaville*

MONTEGO BAY 488 MI
DRY TORTUGAS 70 MI
HONOLULU 4788 MI
SAN FRANCISCO 2550 mi
MINOCQUA 1881 MI
NASSAU 281 MI
NEW ORLEANS 630 MI

The sea does not reward those who are too anxious,
too greedy, or too impatient.
One should lie empty, open, choiceless as a beach—
waiting for a gift from the sea.

—Anne Morrow Lindbergh

At the beach life is different.
Time doesn't move hour to hour,
but mood to moment.
We live by the currents, plan by the tides,
and follow the sun.

—Unknown

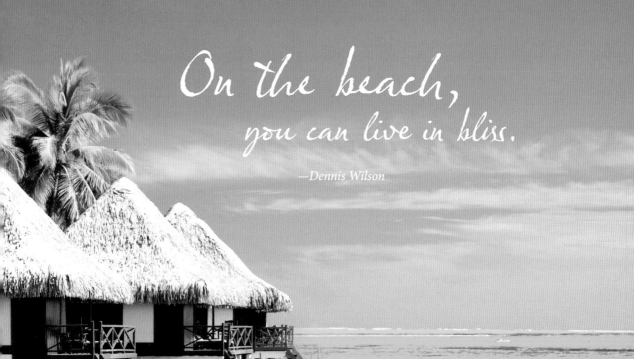

On the beach,
you can live in bliss.

—Dennis Wilson

DON'T

GROW UP TOO QUICKLY, LEST YOU FORGET HOW MUCH

YOU LOVE
THE BEACH.

—Michelle Held

The sea pronounces something,
over and over, in a hoarse whisper;
I cannot quite make it out.

—Annie Dillard

Sea Shell, Sea Shell,
Sing me a song, O Please!
A song of ships, and sailor men,
And parrots, and tropical trees
Of islands lost in the Spanish Main
Which no man may ever find again,
Of fishes and corals under the waves,
And seahorses stabled in great green caves.

—Amy Lowell

In every outthrust headland,
in every curving beach,
in every grain of sand
there is the story of the earth.

—Rachel Carson

COME WALK WITH ME, TAKE OFF YOUR SHOES. LET'S WALK THE BEACH WITH ONLY US AND THE WAVES.

—*Unknown*

When we came within sight of the sea,
the waves on the horizon,
caught at intervals above the rolling abyss,
were like glimpses of another shore
with towers and buildings.

—Charles Dickens

Are you feeling, feeling, feeling like I'm, feeling
Like I'm floating, floating, up above that big blue ocean
Sand beneath our feet, big blue sky above our heads,
No need to keep stressing from our everyday life on our minds
We have got to leave all that behind

—The Avett Brothers, At the Beach

I find myself at the extremity of a long beach.
How gladly does the spirit leap forth,
and suddenly enlarge its sense
of being to the full extent
of the broad, blue, sunny deep!
A greeting and a homage to the Sea!

—Nathaniel Hawthorne

PIÑA COLADA

3 OUNCES LIGHT RUM

3 TABLESPOONS COCONUT CREAM

3 TABLESPOONS CRUSHED PINEAPPLE

RUM RUNNER

1 1/2 OUNCES VODKA

1 1/2 OUNCES PEACH SCHNAPPS

2 OUNCES CRANBERRY JUICE

2 OUNCES ORANGE JUICE

Just as the wave cannot exist for itself,
but is ever a part of the heaving surface of the ocean,
so must I never live my life for itself,
but always in the experience, which is going on around me.

—Albert Schweitzer

The sea has never been friendly to man.
At most it has been the accomplice
of human restlessness.

—Joseph Conrad

My soul is full of longing
For the secret of the sea,
And the heart of the great ocean
Sends a thrilling pulse through me.

—Henry Wadsworth Longfellow

HAPPINESS IS THE SAND BETWEEN MY TOES AND THE SUNBURN ON MY NOSE!

—Unknown

TO THE BEACH